Nursing Mathematics Review

Atheen M. Wilson, RN

AD, BA, BS, MA
Formerly CCRN on SICU
University of Minnesota Hospitals and Clinics
Minneapolis, MN

O Pine Press

Title: Nursing Mathematics Review

ISBN: 1450548059
EAN-139781450548052
Primary Category: Mathematics / General
Country of Publication: United States
Language: English
Search Keywords: Mathematics;Nursing;Medication-dosages
Contributors: *Authored by* Atheen M. Wilson RN

O Pine Press

O Pine Press
803 Huron Blvd SE
Mpls, MN 55414

To Rob Kvilhaug

for reminding me to never get rusty!

Contents

Introduction

Probably one of the most difficult issues in learning to become a registered nurse is getting a handle on pharmaceutical mathematics. It's not that it's particularly difficult, but most people tend to be a little math shy, and those two simple words "math test" have undone many a person. Even though nursing math is just basic math, how many of us use what we learned in grade school beyond addition and subtraction for domestic book keeping or in making change? To help overcome both the "freeze" and the "cobwebs," I've taken on the task of writing a simple approach to the problem that should help both the student and those who need to brush up on the process to pass the test for a new facility.

The system I propose here is one I learned years ago, when I myself was in nursing school. Since that time I have found that few new nurses are familiar with it. Often they have been taught "proportions," which is also very helpful, but not what I call the "units system," which helps to set up the problem from the information that's given in the first place and cues the student when there's something "not quite right" with the problem as they've constructed it. This system is so simple and clears up so much of the inherent confusion that I decided to put out this booklet to help others learn it.

Let me say first, however, that the process often seems counter intuitive. Be assured, however, that it is mathematically sound. At points where I think concern will arise, I will try to explain the mathematical principles, the whys and wherefores of that stage of the problem.

Like grade school word problems, nursing math is about knowing what information you have to start with and what you need to find as your solution. Take time at each stage to understand why this process works, because it is the process itself which will come most readily to mind when confronted with a test or a nursing decision.

Learning is very much an active thing. I encourage you to write your own notes in the margins whenever necessary, explaining the meaning of the content in your own terms. That will ensure that you actually have a grasp on the ideas. Again, it will be the underlying principles that will come back to you later when you most need them.

----The best of luck to you all,

Atheen Wilson

I.

For the Nursing Student: the Whys and Wherefores

For the nursing student let me say that 99% of the time a computer or a pharmacist—probably himself using a computer—will do most of the "figuring" for you in actual practice. The days when a nurse dispensed pills from a common container or even an individual pill bottle are long gone as are mixing vasoactive and other intravenous (IV) medications; the last time I remembered doing it this way was 1985, and it was rare even then. While there are IV medications—Lasix for one—that still require the nurse to calculate the volume of the drug given from a vial or ampoule, this is becoming rarer, especially on pediatrics, and may disappear altogether in the future. The concern over possible errors in computation has brought about standardized dosages packaged in a standardized method of administration. Possible contamination of what should be a sterile IV medium is avoided by having the medication mixed by hospital pharmacists or purchased by the hospital pharmacy premixed from pharmaceutical companies. In both these cases, IV bags are mixed under special laminar flow hoods to cut down that risk. The trend in this direction will probably continue and expand.

Note too, technology has been put in place in many cases, removing the burden of complicated mathematics from the nurse's shoulders. In the case of IV medications, particularly vasoactive ones, most hospitals employ pumps for delivery, and these are programmed for patient weight, drug dose, fluid volume, etc., and with this information calculate the flow rate. In some cases these pumps will also communicate with computerized charting programs to keep records of the medication that was delivered. An increase in this type of technology will also probably continue.

Despite all of this, however, most institutions to which you apply for employment will expect you to take a math test as part of their selection process. It's not that you won't get hired if you "screw up"—they may well provide a remedial class and give you the test again—but putting your best foot forward by knowing what you're doing from the beginning is always a good idea. This is especially the case when jobs are hard to find, which they tend to be on a cyclical basis.

Your individual strength as a new graduate in this test taking situation is that you have probably had mathematics—particularly algebra—more recently, so the process of figuring out an unknown will make more sense to you. Use this strength to your best advantage.

II.

For the Experienced Practitioner: How Did that Go Again?

Most experienced nurses already know that the pharmacy, a computer or a computerized pump will do many of their calculations for them, but there are three good reasons to review.

The first of these is that even computers and pharmacists can come up wrong. If your math doesn't agree with your own computer, the pharmacist, or the pump, you need to ferret out why. Especially with drop rates or micrograms per minute, etc., a quick review of what you *should* expect is always a good idea, especially when checking your pumps at the beginning of your shift. This is when most mistakes are discovered, reported and corrected.

The second reason to review is that, should you decide to change facilities, you will almost certainly be expected to take a math test, and for RN's this will probably involve fluid flow rates as well as the usual dosage estimates. While some of the math will be simple, some of it will not be quite so straight forward, especially if you've been relying on a computer or a pharmacist to do the work for you for a while. A math brush up will be helpful, allowing you to face a test with greater confidence.

The third reason is that review keeps the cobwebs out of your infrastructure! At least it does for me!

Your strength is that having done a lot of the figuring in your head with your own system, you may already know the answer to a question. It is, however, the equation that may be what the tester actually wants to see, since it shows that you understood how you came up with the answer you did. That way they know your answer to a similar question on the floor will not be a matter of "accident" or a "good guess." If your paperwork suggests haphazard methods, the tester may not accept it. That being said, however, if your paper equation comes out wrong, you'll also know it and recheck the equation itself for what's not there but should have been. Knowing the answer by your own method creates a check for your paper results.

III.

The System: How it works

The system I propose here is a design for setting up your equations so that you can solve problems more easily and accurately without trying to figure out how everything is related to everything else. Attempting to intuit how all the factors relate can be time consuming and confusing, especially under the pressure of test taking and its attendant constraints.

The present method is based on the cancellation of units in the factors in the left hand of the equation so that the units required by the question are all that remain and match those in the right hand portion. Even before you do the actual math itself, before you even enter in any numbers at all, you will be able to set up *any* problem properly by making certain your equation brings you to the requested units. In fact, it is ultimately knowing the requested units that allows you to work backwards to set up the equation before you perform the math.

For instance, if you want to know how many capsules are required to arrive at a given dosage, the problem should be framed so as to cancel out all units except "capsules" and "dose."

Here is a demonstration:

> Your order reads: Give 300 mg of Dilantin by mouth every evening.
> Your Dilantin comes in 100 mg capsules; how many would you give?

The first thing to do is to write down in mathmatical form all the units information given in the problem.   You know that you have miligrams per capsule and that you need miligrams per dose; these are therefore your left hand factors. The problem also clearly states what you need to know, ie. capsules per dose. This then becomes your right hand factor or answer units. The combined units tell you everything you need to know in order to create your problem in mathematical form:

$$\frac{mg}{capsule} \; x \; \frac{mg}{dose} = \frac{capsules}{dose}$$

Notice that the equation has "capsules" on the top of the answer but on the bottom of the factor containing it on the left. To get capsules per dose, you will have to invert the first left hand factor. Notice also that as the above equation stands, you would get mg^2 in the answer, which you shouldn't and don't. Since this makes no sense, one of the factors must be inverted to cancel the mg in each so that they are absent in the term on the right. Again, having inverted the first factor to place capsule above and miligrams below in this term also cancels the miligrams in both. See below:

$$\frac{capsule}{\cancel{mg}} \; x \; \frac{\cancel{mg}}{dose} = \frac{capsules}{dose}$$

As you can see, this makes much more sense mathematically. Adding in the numbers now lets you solve the actual problem as posed:

$$\frac{1 \, capsule}{\cancel{100 \, mg}} \; x \; \frac{\cancel{300 \, mg}}{1 \, dose} = \frac{3 \, capsules}{1 \, dose}$$

While this problem is comparatively easy and could be solved much faster in your head by simply adding three 100 miligram capsules for a total of 300 miligrams, other prolems are not always as easily solved that way. For one thing, the more factors that have to be introduced into the equation, the more complicated it becomes. This makes solving it in your head more likely to lead to the neglect of a necessary term, let alone mistakes in the actual manipulation of the numbers. Conversion factors, like the number of mililiters in a liter or the number of pounds in a kilogram, are particularly prone to this type of omission.

The following is a breakdown of the method into simple steps that may be more easily remembered and can form the basis of a step by step approach to your actual problems.

IV.

The Breakdown: Step by step

While the procedure above may seem complicated at first, it can actually be broken down into the following practical points.

1. Remember that almost all of the nursing mathematics in this program involve **multiplication and division** rather than addition and subtraction.

2. **Read** the problem to pick out the important **information** and write it down in a **table**, i.e.) dosage, conversion factors if needed, what you want at the end, etc.

3. **Ignore the numbers**. It's really not about numbers at this stage; the whole notion is about ending up with the **right units**. Once you've written down the right equation, the right numbers become a matter of simple multiplication and division. Even this can be relatively pain free if you cancel the numbers in all the factors to their smallest and then multiple what remains.

4. Examine the problem to determine if you're **changing original units** within a category—like milligrams—to something else—like micrograms. Do it first because it's easy to forget it later if you don't. This is also where you need to know what your **conversion factors** are:

$$1000 \; miligrams \leftrightarrow 1 \; kilogram$$

$$2.2 \; pounds \leftrightarrow 1 \; kilogram$$

$$1000 \; micrograms \leftrightarrow 1 \; miligram$$

In cases like this, if the question is looking for *micrograms* in the answer but you leave it in milligrams, the test proctor may count it as "wrong" even though it may be technically right. It's a way to see if you're paying attention to all aspects of the original order. Essentially it's a trick, but the issue is one of care and exactness; examiners want to know if you will miss this kind of thing in the orders of your future patients.

5. Note however that though you put the conversion factor (here for milligrams to micrograms) in right away, it or other factors may have to be **flipped** over later to achieve your primary goal. This is "legal" because one milligram has 1000 micrograms, and 1000 micrograms equals one milligram. If you cancel out the numbers and units in the equation below, the answer is a unit-less 1, which is a mathematician's way of saying that they are **mathematically identical**:

$$\frac{\cancel{1 \; mg}}{\cancel{1000 \; mcgs}} = \frac{\cancel{1000 \; mcgs}}{\cancel{1 \; mg}} = 1$$

6. Next, put down the **information given** you. It'll be a measured dose of something—like milligrams per tablet, milligrams per liter etc.—or a conversion factor of some kind. This is information that you'd find on your vial or pill bottle to put your dose together. These are your **"givens,"** the starting points for your solving the problem.

Without this information, you couldn't answer the question at all. "If you needed to give the patient 75 milligrams and you had a bottle of tablets, how many would you give?" Who knows? You need to know how many milligrams are in one tablet of the medication. That's why the pharmacy disapproves when patients dump their pills into something without dosage information, and why they always use in-house dispensed medications and not the patient's own supply while the patient remains in the hospital—unless the MD writes a specific order that he or she may do otherwise.

7. Once you have all the information set out in your table, **rearrange the ratios** until all units cancel except the ones you want.

Below is an arrangement of units from an initial list of information provided in a given medication problem. Without knowing *anything else* about the list, which ratios would you need to invert in order to cancel out the appropriate units on the left of the equation to arrive at the units given in the answer to the right?

$$\frac{mcg}{mg} \ x \ \frac{cc}{mg} \ x \ \frac{mcg}{dose} = \frac{cc}{dose}$$

Without even knowing the actual details of the medication problem, you can see that you will need to invert the first factor in the equation in order to cancel out milligrams and micrograms to arrive at cubic centimeters per dose:

$$\frac{\mathbf{mg}}{\mathbf{mcg}} \ x \ \frac{cc}{mg} \ x \ \frac{mcg}{dose} = \frac{cc}{dose}$$

No other factors if inverted would have worked out appropriately for this problem as stated. You already had cc on the top and dose on the bottom at the left, which is what you wanted. To cancel out mg and mcg without changing cc/dose in this equation, you could only invert the first factor.

Having performed the required manipulation of the units, you can now cancel them out as below:

$$\frac{\cancel{mg}}{\cancel{mcg}} \; x \; \frac{cc}{\cancel{mg}} \; x \; \frac{\cancel{mcg}}{dose} = \frac{cc}{dose}$$

Take time to notice at this point that you haven't confused yourself with any numbers at all or with any relationships the units might otherwise be presumed to have with one another or with the numbers; you've just inverted your unit ratios until you cancelled out everything except what you wanted in the answer to your problem.   It really is this simple. **DON'T MAKE IT ANY MORE DIFFICULT THAN IT REALLY IS.**

Let's take a look at some other problems to see from whence all this mathematical mystery comes.

V.

Guided Practice Problems: Pulling it all together

We will start simple, illustrating the above points in more tabular form to impress the technique with actual practice.

Problem 1.

> Your order reads: Give Medication A 400 mcg, subcutaneously, one time, now.
> You have a vial of Medication A with 0.4 mg per milliliter. How many milliliters would you give your patient?

Step 1. The information as you would write it down for yourself in a table would look like this:

> Conversion factor? = yes: 1 mg = 1000 mcgs
> Available dose = 0.4 mg in 1 milliliter
> Needed dose = 400 mcg/dose
> Unknown = milliliters/1 dose

In setting up this table, you should think **CAN=U**, with the caveat that the C factor might be in fact multiple in some problems.

Step 2. In simple units and *without manipulation* at this point, this information converts *initially* to the following equation:

$$\frac{mg}{mcg} \; x \; \frac{mg}{ml} \; x \; \frac{mcg}{dose} = \frac{ml}{dose}$$

[C]   [A]   [N]   =   [U]

All of this comes directly from the information that you listed in your table, one entry after the other. If this is not apparent, reexamine the above table and the equation to see why it is true. If necessary explain it to yourself in your own words, writing a note to yourself in the margin at this point.

Step 3. Now invert the proper units so that the left and right hand portions of the equation arrive at the same units:

$$\frac{mg}{mcg} \; x \; \frac{\boldsymbol{ml}}{\boldsymbol{mg}} \; x \; \frac{mcg}{dose} = \frac{ml}{dose}$$

[C]   [A]   [N]   =   [U]

As you notice, the second factor in the equation—here the available dose—has been flipped over to cancel out the milligrams and to put the desired units in the top position of the ratio.

Illogical as it may seem at first, this changes nothing with respect to the factor: the dose is still 0.4 mg in one ml and 1 ml still contains 0.4 mg! All that changes is that now all the units cancel out appropriately.

Step 4. Now put in the numbers you were given as you wrote them down for yourself, remembering that one of your ratios has been inverted.

$$\frac{1 \; mg}{1000 \; mcg} \; x \; \frac{1 \; ml}{0.4 mg} \; x \; \frac{400 mcg}{dose} = \frac{ml}{dose}$$

[C]   [A]   [N]   =   [U]

Step 5. Now, working out the math, the following equation arises:

$$\frac{\cancel{1\,mg}}{\underset{1}{\cancel{1000\,mcg}}} \; x \; \frac{1\,ml}{\underset{1}{\cancel{0.4mg}}} \; x \; \frac{\overset{\overset{1}{\cancel{1000}}}{\cancel{400mcg}}}{dose} = \frac{1\,ml}{1\,dose}$$

$$[C] \qquad\qquad [A] \qquad\qquad [N] \;\; = \;\; [U]$$

Here the numbers cancel themselves out conveniently, leaving very little by way of actual practical math to perform.

Try another problem like this one to make certain you understand what's going on and why.

Problem 2.

> You have a vial of medication reading: Medication B 10 mg in 2 cc.
> You want to give your patient a 100 mcg dose. How many cc's would you need?

Step 1. Your table:

> Conversion factor? = yes, 1000 mcg = 1 mg
> Available dose = 10 mg/2cc
> Needed dose = 100mcg/1 dose
> Unknown = cc/1 dose

Step 2. From the table and *without manipulation*, your equation:

$$\frac{mcg}{mg} \times \frac{mg}{cc} \times \frac{mcg}{dose} = \frac{cc}{dose}$$

$$[C] \qquad [A] \qquad [N] \; = \; [U]$$

Step 3. Inverting factors to ensure that you can cancel out everything but the answer [U] units:

$$\frac{\cancel{mg}}{\cancel{mcg}} \times \frac{cc}{\cancel{mg}} \times \frac{\cancel{mcg}}{dose} = \frac{cc}{dose}$$

$$[C] \qquad [A] \qquad [N] \; = \; [U]$$

All you did here was invert the first two ratios to cancel everything out except cc's/dose.

Step 4. Now putting in the numbers:

$$\frac{1 \, mg}{1000 \, mcg} \times \frac{2 \, cc}{10 \, mg} \times \frac{100 \, mcg}{1 \, dose} = \frac{cc}{dose}$$

$$[C] \qquad\quad [A] \qquad\quad [N] \quad = \quad [U]$$

Step 5. Now working out the math, the following equations arise:

$$\frac{\cancel{1 \, mg}}{\underset{\underset{50}{100}}{\cancel{1000 \, mcg}}} \times \frac{\overset{1}{\cancel{2 \, cc}}}{\underset{1}{\cancel{10 \, mg}}} \times \frac{\overset{\overset{1}{\cancel{10}}}{\cancel{100 \, mcg}}}{1 \, dose} = \frac{cc}{dose}$$

$$\frac{1\ cc}{50}\ x\ \frac{1}{1\ dose} = \frac{1\ cc}{50\ doses}$$

$$= \frac{0.02\ cc}{dose}$$

Try IV fluids next to see that the system works well here also.

Note in passing here, that for practical purposes, cc's and ml's are treated as equal.

Problem 3.

> You have a 1000cc IV bag of D5W with 500 mg of Medication C in it.
> The order reads 100 mg/hour of Medication C IV until infused.
> At what rate would you run this IV?

Step 1. Your table:

> Conversion factor? = no, in this case mg remain mg.
> Available dose = 500 mg in a 1000 cc bag
> Needed dose = 100 mg//hourly dose
> Unknown = cc/hourly dose

Step 2. Arrange the units in an equation:

$$\frac{mg}{cc} \; x \; \frac{mg}{hr} = \frac{cc}{hr}$$
$$[A] \quad\;\; [N] = \;[U]$$

In this case we do without our [C] factor, because we are not changing our units into anything else. Throughout the equation the milligrams and cubic centimeters remain unchanged. Notice, too, that our "dose" is an hour's worth rather than a tablet or a single sq/IM dose. The patient will receive an hour's worth of medication every hour until all 500 mg are given.

Step 3. Invert factors on the right to eliminate all but cc/hr.

$$\boldsymbol{\frac{cc}{mg}} \; x \; \frac{mg}{hr} = \frac{cc}{hr}$$

Step 4. Add the numbers.

$$\frac{1000 \; cc}{500 \; mg} \; x \; \frac{100 \; mg}{1 \; hr} = \frac{cc}{hr}$$

Step 5. Do the math.

$$\frac{1000 \; cc}{\cancel{500 \; mg}_{\,5}} \; x \; \frac{\cancel{100 \; mg}^{\,1}}{1 \; hr} = \frac{\cancel{1000}cc}{\cancel{5} \; hr}$$

$$= \frac{200 \; cc}{hr}$$

For those of you with experience with IV's, you could easily do all of this in your head, but the examiner may want to see on paper how you arrived at your answer. Whether you realize it or not, you probably did something very similar to the above, though so fast and intuitively you might have had some difficulty in telling them exactly how you did do it!

How long would the above IV run? Figure this out with the same method.

Problem 4.

Conversion factor? = no
Available dose = 1000cc/bag
Needed dose = 200cc/hr (taken from the above answer and because all 500 mg of the Rx need to be infused)
Unknown = hours/bag

The first arrangement of the units:

$$\frac{cc}{bag} \; x \; \frac{cc}{hr} = \frac{hr}{bag}$$

Rearranging to cancel units:

$$\frac{cc}{bag} \; x \; \frac{hr}{cc} = \frac{hr}{bag}$$

Adding in the numbers:

$$\frac{1000 \ cc}{1 \ bag} \ x \ \frac{1 \ hr}{200 \ cc} = \frac{hr}{bag}$$

Doing the math:

$$\frac{\overset{5}{\cancel{1000 \ cc}}}{1 \ bag} \ x \ \frac{1 \ hr}{\underset{1}{\cancel{200 \ cc}}} = \frac{5 \ hr}{1 \ bag}$$

Again and obviously, the experienced practitioner will immediately realize that this can be done in one's head far more quickly than by putting it all on paper. In a test setting however, it always helps to show exactly what you're doing so that the proctor can see that you truly understand how the problem is solved. Furthermore, seeing the paperwork come out the same as your intuitive answer is a good cross check. If it's not the same, either the intuitive or the paper answer is wrong, and you need to ascertain "which" and "why" to do well on the exam. Use your strengths to get the best results.

Try the next problem which contains some aspects that may be confusing at first. This problem shows that no matter how complex the question, approaching it methodically will help you do the "hard part"—that of actually setting up the equation—much more easily.

Problem 5.

> You have a 250cc bag of D5W with 50 mg of Medication D in it mixed by the pharmacy.
> The order reads; Give Medication D at 5 mcg per kg per min by central line.
> You have a pump that can only be programmed for cc/hour.
> Your patient weighs 220 pounds.
> At how many cc's per hour would you run this IV?

At first this problem seems horrendous. There are so many factors that need to be added to the equation, but how do they all relate? Do you take the time to figure out all these relationships, or do you apply the method above to determine just that?

What's first?

The table as before:

> <u>C</u>onversion factor? = yes, three:
> $\qquad$ 1 kg = 2.2 pounds for pt 220 pounds
> $\qquad$ 1000 mcg = 1 mg
> $\qquad$ 60 minutes = 1 hr
> <u>A</u>vailable dose = 50mg in 250 cc
> <u>N</u>eeded dose = 5mcg/kg/1 min
> <u>U</u>nknown = cc/hour

The rough equation:

$$\underset{[C]}{\frac{kg}{pd}} \ x \ \underset{[``C'']}{\frac{pd}{1\ (pt)}} \ x \ \underset{[C]}{\frac{mcg}{mg}} \ x \ \underset{[C]}{\frac{min}{hr}} \ x \ \underset{[A]}{\frac{mg}{cc}} \ x \ \underset{[N]}{\frac{mcg/kg}{min}} = \underset{[U]}{\frac{cc}{hr}}$$

Take a few minutes to see why this problem is arranged as it is. The second factor is technically a "conversion" number, that is to say, one patient weighs 220 pounds. To solve the problem, the patient's weight must be converted from the standard to the metric system of measurement. These two conversion factors could be merged by simply making the final weight in kilograms part of your original table, i.e.):

Conversion factor? = yes, 1 kg = 2.2 pds, for patient weighing 220 pounds = 100 kg.

Here, for the sake of completeness, it was added to the equation. Also needed are the conversion factors for milligrams to micrograms. Finally minutes must be converted to hours.

All of these are things for which you need to read the question carefully before working on an equation. That's why the CAN=U method is helpful. It enables you to mine the information given you in order to identify all aspects of the problem before you attempt to solve it. You don't have to think about the equation at all at this point; all you need to do is set out the information provided in the problem.

Now that all the information has been arranged in a tabular form, we can begin the process of inverting ratios until all except cubic centimeters per hour have been cancelled. It's really the same method as it was before; there are simply more factors to enter. This is where the units system is most helpful, since it allows you to set up the equation itself before adding any numerical math to it.

$$\frac{kg}{pd} \; x \; \frac{pd}{1\,(pt)} \; x \; \frac{mg}{mcg} \; x \; \frac{min}{hr} \; x \; \frac{cc}{mg} \; x \; \frac{mcg/kg}{min} \; = \; \frac{cc}{hr}$$

[C]   ["C"]   [C]   [C]   [A]   [N]   [U]

Starting with the simplest issue first, the cc's must be on top in the left hand portion of the equation in order for them to be on top on the right; thus the mg/cc factor must be inverted to cc/mg. Right away math anxiety will probably kick in when you see the mcg/kg/min. "How do I deal with this?" you might ask, and probably should unless you've had your mathematics training fairly recently or remember it well.

First, forget for the moment the entire lower half of all of the ratios and just focus on the math at the top of the ratios. Think of the problem as looking like this:

$$\frac{kg \; x \; pd \; x \; mg \; x \min x \; cc \; x \; mcg}{kg}$$

(Actually the whole equation looks like this:

$$\frac{\dfrac{kg \; x \; pd \; x \; mg \; x \min x \; cc \; x \; mcg}{kg}}{pd \; x \; 1 \; (pt) \; x \; mcg \; x \; hr \; x \; mg \; x \; min} \quad).$$

Right away you can see that kg can be cancelled:

$$\frac{\cancel{kg} \; x \; pd \; x \; mg \; x \min x \; cc \; x \; mcg}{\cancel{kg}}$$

The equation now looks like this:

$$\frac{1 \; x \; pd \; x \; mg \; x \min x \; cc \; x \; mcg}{1}$$

(Or technically like this:

$$\frac{\frac{1 \, x \, pd \, x \, mg \, x \, min \, x \, cc \, x \, mcg}{1}}{pd \, x \, 1 \, (pt) \, x \, mcg \, x \, hr \, x \, mg \, x \, min})$$

Which actually reduces to this:

$$1 \; x \; pd \; x \; mg \; x \min x \; cc \; x \; mcg$$

Returning the lower portion of the ratios makes it look like this:

$$\frac{1}{pd} \; x \; \frac{pd}{1} \; x \; \frac{mg}{mcg} \; x \; \frac{min}{hr} \; x \; \frac{cc}{mg} \; x \; \frac{mcg}{min} = \frac{cc}{hr}$$

Cancelling out the units makes the equation look like this:

$$\frac{1}{\cancel{pd}} \; x \; \frac{\cancel{pd}}{1} \; x \; \frac{\cancel{mg}}{\cancel{mcg}} \; x \; \frac{\cancel{min}}{hr} \; x \; \frac{cc}{\cancel{mg}} \; x \; \frac{\cancel{mcg}}{\cancel{min}} = \frac{cc}{hr}$$

The pounds cancel themselves out as do the minutes, the micrograms, and the milligrams, leaving only cc/hr as the answer.

If this seems complicated, study the equations again to see what has occurred here and why. Write in the margins in your own words what is occurring so that when you review again, it will be more apparent to you why the equation is as it is.

Understanding the underlying mathematical principles will help you set up any complex problem without trying to remember everything you ever knew about medications and the relationships between all the various factors. Remember, you only need to show you understand the principles for the exam. Computers will probably do the work for you on the floor; but they won't be there for the exam.

Entering the numbers for the above problem produces the following equation:

$$\frac{1}{2.2\ pd} \times \frac{220\ pd}{1} \times \frac{1\ mg}{1000\ mcg} \times \frac{60\ min}{1\ hr} \times \frac{250\ cc}{50\ mg} \times \frac{mcg}{min}$$

Cancelling throughout yields the following equations:

$$\frac{1}{\cancel{2.2\ pd}_{\ 1}} \times \frac{\overset{100}{\cancel{220\ pd}}}{1} \times \frac{1\ \cancel{mg}}{\underset{4}{\cancel{1000\ mcg}}} \times \frac{60\ \cancel{min}}{1\ hr} \times \frac{\overset{1}{\cancel{250}}\ cc}{\underset{10}{\cancel{50\ mg}}} \times \frac{\overset{1}{\cancel{5}}\ \cancel{mcg}}{1\ \cancel{min}} =$$

$$\frac{1}{1} \times \frac{100}{1} \times \frac{1}{4} \times \frac{60}{1\ hr} \times \frac{1\ cc}{10} \times \frac{1}{1} = \frac{cc}{hr}$$

$$\frac{\overset{25}{\cancel{100}}}{1} \times \frac{1}{4} \times \frac{\overset{6}{\cancel{60}}}{1\ hr} \times \frac{1\ cc}{\underset{1}{\cancel{10}}} = \frac{150\ cc}{hr}$$

Review the preceding problem carefully to see what was cancelled and how the ultimate answer transpired. When you feel confident with your results, try the next problem.

With the following examples, perform each step of the problem before checking the guided portion to see if you are correct in your assumptions.

<u>Problem 6.</u>

You have a one liter bag of normal saline with 500 mg of Medication E in it sent by the pharmacy.
The order reads: Give Medication E at 10 mcg per kilogram per hour by central line. Your patient weighs 440 pounds.
At what rate would you run this infusion?

Step 1.

Did you remember what step one was? If not, look at the preceding examples. Did you remember the mnemonic CAN=U?

Did your table look something like this?

Conversion factor? = yes, three:
       1 kg/2.2 pounds, for pt 440 pounds   (200 kg)
       1000 mcg/1 mg
       1000cc/1L
Available dose = 500mg/1L
Needed dose = 10 mcg/kg/hr
Unknown = cc/hour

If not, write down in your own words in the empty space on the previous page any memos that might aid you next time. Also explain to yourself what was absent and why it was significant.

Step 2.

Can you recall what your next step should be? If not, review section IV, The Breakdown: Step by step, and look back at some of the previous examples.

Did your initial equation should look something like this?

$$\frac{kg}{pd} \; x \; \frac{pd}{1 \, (pt)} \; x \; \frac{mcg}{mg} \; x \; \frac{cc}{L} \; x \; \frac{mg}{L} \; x \; \frac{mcg/kg}{hr} = \frac{cc}{hr}$$

If not, study the previous examples to see why it was different.

Perhaps you jumped a step and inverted the appropriate factors on the left to cancel all but the units expected on the right. This is an acceptable leap, so long as it was a conscious one. If this is not the explanation, write in your own words in the empty space on the previous page what errors occurred, what measures you might take to correct them, and how to remember this step. Explain to yourself why it is important to the procedure. This will help you if you repeat the review.

Step 3.

Do you recall the third step? If not, check back to the previous examples and see section IV, The Breakdown: Step by step. Guided problems have the inherent limitation of pulling the reader along passively. This one is designed to test your memory and encourage you to work with it, which makes the learning a more active behavior.

In inverting to cancel out all but the desired units, did your equation look like this one?

$$\frac{kg}{pd} \; x \; \frac{pd}{1} \; x \; \frac{\boldsymbol{mg}}{\boldsymbol{mcg}} \; x \; \frac{cc}{L} \; x \; \frac{\boldsymbol{L}}{\boldsymbol{mg}} \; x \; \frac{mcg/kg}{hr} = \frac{cc}{hr}$$

$$[C] \quad [C] \quad [C] \quad [C] \quad [A] \quad [N] \quad [U]$$

If not, study the previous examples to see where you might have gone wrong. Here again, make notes in the space remaining on the previous page to explain in your own words how you might remember the procedure at this step and why it's important.

Step 4.

What is step 4?

Did you remember what step 4 was? With the numbers entered, did your equation look like this?

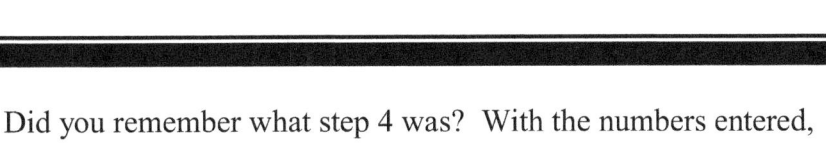

$$\frac{1\ kg}{2.2\ pd} \quad x \quad \frac{440\ pd}{1} \quad x \quad \frac{1\ mg}{1000\ mcg} \quad x \quad \frac{1000\ cc}{1\ L} \quad x \quad \frac{1\ L}{500\ mg} \quad x \quad \frac{10\frac{mcg}{kg}}{1\ hr}$$

$$= \frac{cc}{hr}$$

Where did these numbers come from?

Go on to step 5. Can you remember what to do here?

Step 5.

Did math anxiety make you panic, or did you remember what to do here?

$$\frac{1\ kg}{\underset{1}{2.2\ pd}} \ x\ \frac{\overset{4}{\overset{}{\cancel{200}}}}{1} \ x\ \frac{1\ mg}{\underset{1}{\cancel{1000\ mcg}}} \ x\ \frac{\overset{1}{\cancel{1000}}\ cc}{1\ \cancel{L}} \ x\ \frac{1\ \cancel{L}}{\underset{\underset{1}{50}}{500\ mg}} \ x\ \frac{\overset{1}{\cancel{10}}\ \frac{mcg}{kg}}{1\ hr}$$

$$= \frac{4\ cc}{hr}$$

Notice that all units except cc/hr have been excluded from the equation at this point and only the numbers need to be dealt with. Did you understand how the kilograms and micrograms were cancelled? If not, re-read pages 24-26 above for review.

As you can see, what started out a pretty daunting problem reduced itself by stages to some very simple math. Almost all "story problems" are like this.

For the new student nurse, this may all see like preaching to the choir; you've probably had math more recently, so the rules are more immediately available to you. For those who've been away from formal math for a while, much of this will seem vaguely familiar, and it is my hope that this review will help reestablish old concepts more firmly in your mind again.

If the principles still seem vague to you, study the above two examples again until you can understand the principles underlying them. Understanding the "whys" will help you set up your problems with confidence.

Moving on now to drop rates, you will see that the system works equally well for these problems.

Problem 7.

As an introductory note here, I would like to point out that while very few sites actually depend upon a gravity drip for their IV's, they probably will have drip (gtt) rate problems on their tests. As those with nursing experience and students already familiar with health care environments already know, most IV's are managed by pumps; this ensures dose accuracy, delivery safety, and volume control. In the "Olden Days," however, and in some settings for maintenance IV's even now, the drop rate was/is key to managing your IVs by the gravity method. You simply marked your bag or glass bottle for the number of cc's/hr desired, counted the number of drops per minute desired for a given hourly rate, and closed or opened your IV tubing clamp or regulator until you arrived at the desired rate.

Every tubing package indicates its drop factor. Most regular IV tubing—the macro drop—is 15 drops/ml, while micro drop or mini drip tubing is 60 drops/ml. There are other types of tubing that can have different drop rates however—blood product tubing, inter-lipid tubing, and that used for nitroglycerin, for instance, can all have different gauges—so it is still best to check—though these are also the types of tubing most likely to be managed with pumps.

Generally speaking and given the trend toward rigidly standardized equipment, it seems highly likely that all tubing will ultimately be 15 or 60 drops/ml and that all IVs, without exception, will be managed by pumps in the foreseeable future.

The process for setting up your problem is the same here as before.

The order reads: NS IV to run at 150cc/hr.
With regular tubing, at how many drops per minute would this IV run?

What would your table look like?

Did it look like this one?

> $\underline{C}$onversion factor? = yes, 1 hour = 60 minutes
> $\underline{A}$vailable = 15 gtt/ml
> $\underline{N}$eeded dose = 150ml/hr
> $\underline{U}$nknown = drops (gtts)/min

Did you note that cc and ml are equal? (See the note on p. 25).

What is step two?

Is this what you got?

$$\frac{hr}{min} \; x \; \frac{gtts}{ml} \; x \; \frac{ml}{hr} = \frac{gtts}{min}$$

$$[C] \qquad [A] \qquad [N] \; = \; [U]$$

With this illustration the "available" [A] factor is now tubing and its drop factor. Here the inner gauge of the tubing permits 15 drops of fluid to pass through it for each milliliter of IV fluid.

      If you look closely at the above equation, you will see that your units are arranged properly for cancellation of terms directly from your table, so you do not have to invert any of the factors.

Adding in the numbers, what does your equation look like?

Is this the result you got?

$$\frac{1\ hr}{60\ min} \ x \ \frac{15\ gtts}{1\ ml} \ x \ \frac{150\ ml}{1\ hr} = \frac{gtts}{min}$$

Cancelling numbers and units, what is your result?

Is this what you arrived at after all cancellation was completed?

$$\frac{1 \ \cancel{hr}}{\cancel{60} \ min} \ x \ \frac{\overset{1}{\cancel{15}} \ gtts}{1 \ \cancel{ml}} \ x \ \frac{150 \ \cancel{ml}}{1 \ \cancel{hr}} = \frac{150 \ gtts}{4 \ min}$$

$$= \frac{37^+ \ gtts}{1 \ min}$$

Obviously your answer of 37.5 gtts/min is not a workable rate. Nor is a gravity drip as exact as a pump. The effect of gravity on the bag will change as the bag is emptied of its contents, or if it is raised higher or lower from the ground, or if the patient is active or inactive—which is why pumps have become so popular. In actual practice the rate could be anywhere from 36-40 drops per minute and still achieve the desired rate per hour; however you would have to repeatedly check your IV to make certain it was on time, speeding or slowing it accordingly, to achieve the ordered rate— which is another reason pumps are popular, cost of nursing time.

Go on to the next problem with mini drip tubing to see why it was, and is, so popular. Again, try to do your work before checking the guided problem to see if you understand the principles.

Problem 8.

You have a 1000cc bag of D5W to be run by gravity drip at 50cc/hr.
You have minidrip tubing for this purpose.
At what drip rate would you set it?

What's the first step?

Did you create a table like this one?

Conversion factor? = yes, 1 hour = 60 minutes
Available = 15 gtt/cc
Needed dose = 50cc/hr
Unknown = drops (gtts)/min

What now?

Did your initial equation look like this?
If you substituted milliliters for cubic centimeters that would also be correct, but in that case, both factors with these units in them should be the same to prevent confusion.

$$\frac{hr}{min} \; x \; \frac{gtts}{cc} \; x \; \frac{cc}{hr} = \frac{gtts}{min}$$

What's next?

Did you notice that the units already cancel themselves out appropriately? If you skipped the inversion step and proceeded to the addition of the numbers, good for you! In that case, did your final equation look like this?

$$\frac{1\ hr}{60\ min} \; x \; \frac{60\ gtts}{1\ cc} \; x \; \frac{50\ cc}{1\ hr} = \frac{50\ gtts}{min}$$

Again you can see with this illustration that your "available" factor is now tubing. In this instance, the inner gauge of the tubing permits 60 drops of fluid to pass through it, rather than 15, for each milliliter of IV fluid, so the drop factor is 60.

Now do the math of the equation.

Did your equation work out like this one?

$$\frac{1 \, \cancel{hr}}{\cancel{60} \, min} \; x \; \frac{\overset{1}{\cancel{60}} \, gtts}{1 \, \cancel{cc}} \; x \; \frac{50 \, \cancel{cc}}{1 \, \cancel{hr}} = \frac{50 \, gtts}{min}$$

Because the drop rate cancels out the minutes per hour, the drop rate is the same as the volume rate. This is always the case, no matter how many cc/hr you want your IV to run. Without even doing any math at all, you would know that if you hang a micro drip tube, the cc/hr is also drip rate per hour. The caveat here is that counting a volume of 150cc/hr becomes problematic; which is why most micro drip tubing is used for lower volume rates, such as those used on pediatrics.

VI.

Important Memory Details: What do I actually have to memorize?

With the method herein described, the test taker need only memorize common metric conversions like pounds per kilogram, possibly ounces to milliliters, etc., and the metric units equal to one another.

For your convenience, I include the following table:

---

1 ounce ⇔ 30 cc (or ml)
1 liter ⇔ 1000 cc (or ml)
1 kilogram ⇔ 2.2 pounds
1 kilogram ⇔ 1000 grams
1 gram ⇔ 1000 milligrams
1 milligram ⇔ 1000 micrograms

Notes: 1) Although technically not identical, for practical use in health care, cubic centimeters (cc) are the same as milliliters (ml) and are used interchangeably by most institutions.

2) While milliequivalents (mEq) are frequently used for some medications, the milliequivalent is a **chemical unit of measurement**; it is not part of the metric system. Some common and important--and somewhat dangerous--medications come in mEq, and you should be aware of how they are measured and used. Potassium is a common example.

3) "Units" are also used in dosages but are also not really metric; here the term makes a reference to the amount needed to cause an effective reaction in a test animal, ie) mouse units. Insulin, heparin, bee venom, snake venom and some other less common substances are--or were--measured in this way.

VII.

Practice Problems

These practice problems are not really different from those you have already seen in the context of the above discussion. At this point you should be able to apply the method to the new information and come up with the correct answer to the questions.

Ample space has been provided for your use for each problem. After you have completed the questions check the answers and explanations provided on page 57 and following, below. If you've had difficulty, re-read sections III (p. 12 ff) and IV (p. 14 ff). Write notes on each problem that might help you to review later.

Note especially here that the issue is not to decide if the rate of infusion, a dosage or other issues having to do with actual medications are "appropriate" Those issues will doubtlessly be dealt with in other facets of the general exams given by the employer. Here the interest is *strictly* in setting up an equation and arriving at the correct answer.

Problem 1.

Medication A comes in a vial labeled 500 mcg in 2 cc.
The order is for 0.25 mg of this medication.
How many cc's would you give?

Solution and answer on p. 71.

Problem 2.

Medication B comes in liquid form at 30 mg in 5 ml.
Your order is to give your patient 150 mg via his gastric tube.
How many milliliters would you give this patient?

Solution and answer on p. 73.

Problem 3.

Medication C comes in tablet form of 25 mg per tablet.
The order is for 50 mg of medication B.
How many tablets of this drug would you give the patient?

Solution and answer on p. 74

Problem 4.

The order is for 15 mEq of drug D by nasal gastric tube.
The medication arrives from the pharmacy in a 100cc bottle labeled 20 mEq D in 15 cc.
How many cc's would you put down the NGT for this patient?

Solution and answer on p. 75.

Problem 5.

Drug E comes in 200 mg in 1cc in a 10cc vial.
The order reads, 500 mg of medication E.
How many cc would you give of this drug?

Solution and answer on p. 76.

Problem 6.

You must give a patient 12.5 mg of medication F.
When it arrives from the pharmacy, it comes in 2.5 mg tablets of F.
How many tablets of F will you give?

Solution and answer on p. 77.

Problem 7.

The order is for antibiotic G, 300,000 units by piggy back IV.
The vial directs you to add 4.2 ml of NS to make a concentration
of 3 million units in 5cc.
How many cc's will you mix into your piggy back IV bag?

Solution and answer on p. 78.

Problem 8.

Your order is for 1 liter of D5 1/2NS with 20 mEq KCL to run over 8 hours.
What is your hourly rate for this IV?

Solution and answer on p. 80.

Problem 9.

Using macro drip tubing to run the above IV by gravity, what will your drip rate be?

Solution and answer on p. 81.

Problem 10.

If you used micro drip tubing for it, what would your drop rate be?

Solution and answer on p. 82.

Problem 11.

Your order is for antibiotic K, 1000 mg diluted in 250 cc of NS, to run piggyback over 2 hours.
 What would your drip rate be using macro tubing?

Solution and answer on p. 83.

Problem 12.

What would it be using micro tubing?

Solution and answer on p. 84.

Problem 13.

Medication M 500 mg is diluted in 500 cc.
 Your patient weighs 44 pounds.
 You are requested to deliver the drug at 1 mg/kg/hour by IV.
 At what rate would you run this IV?

Solution and answer on p. 85.

Problem 14.

Your patient weighs 25 kg and is to receive 5mcg/kg/min of drug N for eight hours.
Drug N comes diluted at 500 mg in 250 cc of NS.
At what rate would you run this IV?

Solution and answer on p. 86.

Problem 15.

The patient weights 220 pounds.
His doctor orders medication O to be infused at 5 mcg/kg/min.
The IV arrives from the pharmacy with 100 mg O in 250cc NS.
At how many cc/hr would you run this IV?

Solution and answer on p. 88.

Problem 16.

A new order arrives stating that O should be reduced to 3 mcg/kg/hour if the patient's blood pressure drops below 140/80. It does.

 At how many cc/hr would the IV now run?

Solution and answer on p. 90.

Problem 17.

A patient weighing 150 kg is to have an IV of 50 mg Q in 250 cc NS to run at 0.5 mcg/kg/min. At how many cc/hour would this IV run?

Solution and answer on p. 92.

Problem 18.

A new order arrives stating that Q should run at 1.0 mcg/kg/min if the patient's intracranial pressure rises above a reading of 12. It rises.
Now how many cc/hr would the IV run?

Solution and answer on p. 93.

VIII.

Answers and Explanations to Practice Problems

These are the answers to problems 1-18 above. They are intended to give you further experience with the units method of designing equations in fulfillment of word problems presented in medication tests given by employers and others.

Problem 1.

> Medication A comes in an vial labeled 500 mcg in 2 cc.
> The order is for 0.25 mg of this medication.
> How many cc's would you give?

> Conversion factor? = yes, 1000 mcg = 1 milligram
> Available dose = 500 mcg/2 cc
> Needed dose = 0.25 mg/dose
> Unknown = cc/dose

$$\frac{mcg}{mg} \; x \; \frac{mcg}{cc} \; x \; \frac{mg}{dose} = \frac{cc}{dose}$$
$$[C] \qquad [A] \qquad [N] \; = \; [U]$$

$$\frac{mcg}{mg} \; x \; \frac{\boldsymbol{cc}}{\boldsymbol{mcg}} \; x \; \frac{mg}{dose} = \frac{cc}{dose}$$
$$[C] \qquad [A] \qquad [N] \; = \; [U]$$

$$\frac{1000\ mcg}{1\ mg} \quad x \quad \frac{2\ cc}{500\ mcg} \quad x \quad \frac{0.25\ mg}{1\ dose} \quad = \quad \frac{cc}{dose}$$

$$\quad [C] \qquad\qquad [A] \qquad\qquad [N] \qquad = \qquad [U]$$

$$\frac{\overset{2}{\cancel{1000\ mcg}}}{1\ mg} \quad x \quad \frac{2\ cc}{\underset{1}{\cancel{500\ mcg}}} \quad x \quad \frac{0.25\ mg}{1\ dose} \quad = \quad \frac{1\ cc}{dose}$$

$$\quad [C] \qquad\qquad [A] \qquad\qquad [N] \qquad = \qquad [U]$$

Problem 2.

Medication B comes in liquid form at 30 mg in 5 ml.
Your order is to give your patient 150 mg via his gastric tube.
How many milliliters would you give this patient?

Conversion factor? = no
Available dose = 30mg in 5ml
Needed dose = 150 mg/dose
Unknown = ml/dose

$$\frac{mg}{ml} \; x \; \frac{mg}{dose} = \frac{ml}{dose}$$
$$[A] \qquad [N] \; = \; [U]$$

$$\frac{\mathbf{ml}}{\mathbf{mg}} \; x \; \frac{mg}{dose} = \frac{ml}{dose}$$
$$[A] \qquad [N] \; = \; [U]$$

$$\frac{5 \; ml}{30 \; mg} \; x \; \frac{150 \; mg}{1 \; dose} = \frac{ml}{dose}$$
$$[A] \qquad\quad [N] \; = \; [U]$$

$$\frac{5 \; ml}{\cancel{30 \; mg}} \; x \; \frac{\overset{5}{\cancel{150 \; mg}}}{1 \; dose} = \frac{25 \; ml}{1 \; dose}$$
$$[A] \qquad\quad [N] \; = \; [U]$$

sProblem 3.

Medication C comes in tablet form of 25 mg per tablet.
The order is for 50 mg of medication B.
How many tablets of this drug would you give the patient?

Conversion factor? = no
Available dose = 25mg/tablet
Needed dose = 50 mg/dose
Unknown = tablets/dose

$$\frac{mg}{tablet} \; x \; \frac{mg}{dose} = \frac{tablets}{dose}$$
$$[A] \qquad [N] \; = \; [U]$$

$$\frac{\textbf{tablet}}{\textbf{mg}} \; x \; \frac{mg}{dose} = \frac{tablets}{dose}$$
$$[A] \qquad [N] \; = \; [U]$$

$$\frac{1 \; tablet}{25 \; mg} \; x \; \frac{50 \; mg}{1 \; dose} = \frac{tablets}{dose}$$
$$[A] \qquad [N] \; = \; [U]$$

$$\frac{1 \; tablet}{\cancel{25 \; mg}_{1}} \; x \; \frac{\cancel{50 \; mg}^{2}}{1 \; dose} = \frac{2 \; tablets}{1 \; dose}$$
$$[A] \qquad [N] \; = \; [U]$$

Problem 4.

The order is for 15 mEq of drug D by nasal gastric tube.
The medication arrives from the pharmacy in a100 cc bottle
labeled 20 mEq D in 15 cc.
How many cc's would you put down the NGT for this patient?

Conversion factor? = no
Available dose =   20 mEq/15cc
Needed dose =   15 mEq/dose
Unknown =   cc/dose

$$\frac{mEq}{cc} \quad x \quad \frac{mEq}{dose} = \frac{cc}{dose}$$
$$[A] \qquad [N] \quad = \quad [U]$$

$$\frac{cc}{mEq} \quad x \quad \frac{mEq}{dose} = \frac{cc}{dose}$$
$$[A] \qquad [N] \quad = \quad [U]$$

$$\frac{15\ cc}{20\ mEq} \quad x \quad \frac{15\ mEq}{1\ dose} = \frac{cc}{dose}$$
$$[A] \qquad\quad [N] \quad = \quad [U]$$

$$\frac{\overset{3}{\cancel{15}}\ cc}{\underset{4}{\cancel{20\ mEq}}} \quad x \quad \frac{15\ \cancel{mEq}}{1\ dose} = \frac{45\ cc}{4\ dose} = \frac{11.25\ cc}{1\ dose}$$
$$[A] \qquad\qquad [N] \quad = \qquad\qquad\qquad [U]$$

Problem 5.

Drug E comes in 200 mg in 1cc in a 10cc vial.
The order reads, 500 mg of medication E.
How many cc would you give of this drug?

Conversion factor? = no
Available dose =   200mg/1cc in a 10cc vial
Needed dose = 500 mg/dose
Unknown =   cc/dose

$$\frac{mg}{cc} \quad x \quad \frac{mg}{dose} = \frac{cc}{dose}$$
$$[A] \qquad [N] \quad = \quad [U]$$

$$\frac{\mathbf{cc}}{\mathbf{mg}} \quad x \quad \frac{mg}{dose} = \frac{cc}{dose}$$
$$[A] \qquad [N] \quad = \quad [U]$$

$$\frac{1\ cc}{200\ mg} \quad x \quad \frac{500\ mg}{1\ dose} = \frac{cc}{dose}$$
$$[A] \qquad\quad [N] \quad = \quad [U]$$

$$\frac{1\ cc}{\cancel{200\ mg}_1} \quad x \quad \frac{\overset{2,5}{\cancel{500\ mg}}}{1\ dose} = \frac{2.5\ cc}{1\ dose}$$
$$[A] \qquad\quad\ [N] \quad = \quad [U]$$

Problem 6.

You must give a patient 12.5 mg of medication F.
When it arrives from the pharmacy, it comes in 2.5 mg tablets of F.
How many tablets of F will you give?

Conversion factor? = no
Available dose = 2.5 mg/tablet
Needed dose = 12.5 mg/dose
Unknown = tablets/dose

$$\frac{mg}{tablet} \; x \; \frac{mg}{dose} = \frac{tablets}{dose}$$
$$[A] \qquad [N] \; = \quad [U]$$

$$\frac{\boldsymbol{tablet}}{\boldsymbol{mg}} \; x \; \frac{mg}{dose} = \frac{tablets}{dose}$$
$$[A] \qquad [N] \; = \quad [U]$$

$$\frac{1 \; tablet}{2.5 \; mg} \; x \; \frac{12.5 \; mg}{1 \; dose} = \frac{tablets}{dose}$$
$$[A] \qquad\quad [N] \; = \quad [U]$$

$$\frac{1 \; tablet}{\cancel{2.5 \; mg}} \; x \; \frac{\overset{5}{\cancel{12.5 \; mg}}}{1 \; dose} = \frac{5 \; tablets}{dose}$$
$$[A]\!\!\!\underset{1}{} \qquad\quad [N] \; = \quad [U]$$

Problem 7.

The order is for antibiotic G, 300,000 units by piggy back IV.
The vial directs you to add 4.2 ml of NS to make a concentration
of 3 million units in 5cc.
How many cc's will you mix into your piggy back IV bag?

Conversion factor? = no
Available dose = 3,000,000U/5cc
Needed dose = 300,000 U/1 dose
Unknown = cc/dose

$$\underset{[A]}{\frac{units}{cc}} \; x \; \underset{[N]}{\frac{units}{dose}} = \underset{[U]}{\frac{cc}{dose}}$$

$$\underset{[A]}{\frac{\mathbf{cc}}{\mathbf{units}}} \; x \; \underset{[N]}{\frac{units}{dose}} = \underset{[U]}{\frac{cc}{dose}}$$

$$\underset{[A]}{\frac{5 \; cc}{3,000,000 \; units}} \; x \; \underset{[N]}{\frac{300,000 \; units}{1 \; dose}} = \underset{[U]}{\frac{cc}{dose}}$$

$$\underset{\substack{[A] \\ 10}}{\frac{5 \; cc}{\cancel{3,000,000 \; units}}} \; x \; \underset{[N]}{\frac{\overset{1}{\cancel{300,000 \; units}}}{1 \; dose}} = \underset{[U]}{\frac{cc}{dose}}$$

$$\frac{\overset{1}{\cancel{5}}\,cc}{\underset{[A]\atop 2}{\cancel{10\ units}}} \quad x \quad \frac{1}{1\ dose} \quad = \quad \frac{0.5\ cc}{dose}$$
[A]   [N]   =   [U]

Problem 8.

Your order is for 1 liter of D5 1/2NS with 20 mEq KCL to run over 8 hours.
What is your hourly rate for this IV?

Conversion factor? = yes, 1 liter = 1000 cc
Available dose = 1 liter/bag
Needed dose = 8 hours/bag
Unknown = cc/hr

$$\frac{L}{cc} \; x \; \frac{L}{bag} \; x \; \frac{hr}{bag} = \frac{cc}{hr}$$

$$\frac{\mathbf{cc}}{\mathbf{L}} \; x \; \frac{L}{bag} \; x \; \frac{\mathbf{bag}}{\mathbf{hr}} = \frac{cc}{hr}$$

$$\frac{1000 \; cc}{1L} \; x \; \frac{1 \; L}{1 \; bag} \; x \; \frac{1 \; bag}{8 \; hr} = \frac{cc}{hr}$$

$$\frac{\overset{125}{\cancel{1000}} \, cc}{\cancel{1L}} \; x \; \frac{\cancel{1 \, L}}{\cancel{1 \, bag}} \; x \; \frac{\cancel{1 \, bag}}{\underset{1}{\cancel{8}} \; hr} = \frac{125 \; cc}{hr}$$

Problem 9.

Using macro drip tubing to run the above IV by gravity, what will your drip rate be?

Conversion factor? = yes, 1 hour = 60 minutes
Available dose = 15 drops/cc
Needed dose = 125 cc/hr
Unknown = drops/min

$$\frac{hr}{min} \; x \; \frac{drops}{cc} \; x \; \frac{cc}{hr} = \frac{drops}{min}$$

$$\frac{1 \; hr}{60 \; min} \; x \; \frac{15 \; drops}{1 \; cc} \; x \; \frac{125 \; cc}{1 \; hr} = \frac{drops}{min}$$

$$\frac{\cancel{1 \; hr}}{\underset{4}{\cancel{60}} \; min} \; x \; \frac{\overset{1}{\cancel{15}} \; drops}{\cancel{1 \; cc}} \; x \; \frac{125 \; \cancel{cc}}{\cancel{1 \; hr}} = \frac{drops}{min}$$

$$\frac{1}{\underset{1}{\cancel{4} \; min}} \; x \; \frac{1 \; drops}{1} \; x \; \frac{\overset{31.2}{\cancel{125}}}{1} = \frac{31.2 \; drops}{min}$$

Problem 10.

If you used micro drip tubing for it, what would your drop rate be?

Drop rate for micro-drip tubing is the same as cc/hr. The answer is 125 drops/min. If this does not make sense to you, review page 36 and following pages.

Problem 11.

Your order is for antibiotic K, 1000 mg diluted in 250 cc of NS, to run piggyback over 2 hours.
 What would your drip rate be using macro tubing?

Conversion factor? = yes, 1 hour = 60 minutes
Available dose = 15 drops/cc
Needed dose = 250cc/2 hours
Unknown = drops/min

$$\frac{hr}{min} \; x \; \frac{drops}{cc} \; x \; \frac{cc}{hr} = \frac{drops}{min}$$

$$\frac{1\,hr}{60\,min} \; x \; \frac{15\,drops}{1\,cc} \; x \; \frac{250\,cc}{2\,hr} = \frac{drops}{min}$$

$$\frac{\cancel{1\,hr}}{\underset{4}{\cancel{60}\,min}} \; x \; \frac{\overset{1}{\cancel{15}\,drops}}{\cancel{1\,cc}} \; x \; \frac{250\,\cancel{cc}}{2\,\cancel{hr}} = \frac{drops}{min}$$

$$\frac{1}{\underset{1}{\cancel{4}\,min}} \; x \; \frac{1\,drops}{1} \; x \; \frac{\overset{62.5}{\cancel{250}}}{2} = \frac{drops}{min}$$

$$\frac{1}{1\,min} \; x \; \frac{1\,drops}{1} \; x \; \frac{\overset{31.25}{\cancel{62.5}}}{\underset{1}{\cancel{2}}} = \frac{31^{+}drops}{min}$$

Problem 12.

What would it be using micro tubing?

Drop rate for micro-drip tubing is the same as cc/hr. The answer is 125 drops/min. [Did you notice that the IV was 250cc/2hours?]

Problem 13.

Medication M 500 mg is diluted in 500 cc.
Your patient weighs 44 pounds.
You are requested to deliver the drug at 1 mg/kg/hour by IV.
At what rate would you run this IV?

Conversion factor? = yes, 1 kg =2.2 pounds for a pt weighing 44 pounds.
Available dose = 500 mg M/500cc
Needed dose = 1 mg/kg/hour
Unknown = cc/hr

$$\frac{kg}{pd} \; x \; \frac{pd}{1} \; x \; \frac{mg}{cc} \; x \; \frac{mg/kg}{hr} = \frac{cc}{hr}$$

$$\frac{kg}{pd} \; x \; \frac{pd}{1} \; x \; \frac{\mathbf{cc}}{\mathbf{mg}} \; x \; \frac{mg/kg}{hr} = \frac{cc}{hr}$$

$$\frac{1 \, kg}{2.2 \, pd} \; x \; \frac{44 \, pd}{1} \; x \; \frac{500 \, cc}{500 \, mg} \; x \; \frac{1 \, mg/kg}{1 \, hr} = \frac{cc}{hr}$$

$$\frac{1 \, \cancel{kg}}{\cancel{2.2 \, pd}} \; x \; \frac{\overset{20}{\cancel{44 \, pd}}}{1} \; x \; \frac{\overset{1}{\cancel{500 \, cc}}}{\underset{1}{\cancel{500 \, mg}}} \; x \; \frac{1 \, \cancel{mg/kg}}{1 \, hr} = \frac{20 \, cc}{hr}$$

Problem 14.

Your patient weighs 25 kg and is to receive 5mcg/kg/min of drug N for eight hours.
Drug N comes diluted at 500 mg in 250 cc of NS.
At what rate would you run this IV?

Conversion factor? = yes, two: 60 min = 1 hr and 1000mcg = 1mg
Available dose = 500 mg N in 250cc
Needed dose = 5 mcg/kg/min for pt 25 kg
Unknown = cc/hr

$$\frac{min}{hr} \; x \; \frac{mcg}{mg} \; x \; \frac{mg}{cc} \; x \; \frac{mcg/kg}{min} \; x \; \frac{kg}{1\,(pt)} = \frac{cc}{hr}$$

Did you forget to factor in your patient's weight?

$$\frac{min}{hr} \; x \; \frac{\mathbf{mg}}{\mathbf{mcg}} \; x \; \frac{\mathbf{cc}}{\mathbf{mg}} \; x \; \frac{mcg/kg}{min} \; x \; \frac{kg}{1\,(pt)} = \frac{cc}{hr}$$

$$\frac{60\,min}{1\,hr} \; x \; \frac{1\,mg}{1000\,mcg} \; x \; \frac{250\,cc}{500\,mg} \; x \; \frac{5\,mcg/kg}{1\,min} \; x \; \frac{25\,kg}{1\,(pt)} = \frac{cc}{hr}$$

$$\frac{60\,\cancel{min}}{1\,hr} \; x \; \frac{1\,\cancel{mg}}{\underset{4}{\cancel{1000\,mcg}}} \; x \; \frac{\overset{1}{\cancel{250\,cc}}}{\underset{100}{\cancel{500\,mg}}} \; x \; \frac{\overset{1}{\cancel{5\,mcg/kg}}}{\cancel{1\,min}} \; x \; \frac{25\,\cancel{kg}}{1\,(pt)} = \frac{cc}{hr}$$

$$\frac{\overset{15}{\cancel{60}}}{1 \ hr} \ x \ \frac{1}{\underset{1}{4}} \ x \ \frac{1 \ cc}{\underset{4}{\cancel{100}}} \ x \ \frac{\overset{1}{\cancel{25}}}{1 \ (pt)} = \frac{cc}{hr}$$

$$\frac{15 \ cc}{4 \ hr} = \frac{3.75 \ cc}{1 \ hr}$$

Problem 15.

The patient weights 220 pounds.
His doctor orders medication O to be infused at 5 mcg/kg/min.
The IV arrives from the pharmacy with 100 mg O in 250cc NS.
At how many cc/hr would you run this IV?

Conversion factor? = yes, 1 kg = 2.2 pds, 1 mg = 1000 mcg, 1 hr =
60 min
Available dose = 100 mg of O in 250 cc
Needed dose = 5 mcg/kg/min for pt 220 pounds.
Unknown = cc/hr

$$\frac{kg}{pd} \; x \; \frac{mg}{mcg} \; x \; \frac{min}{hr} \; x \; \frac{mg}{cc} \; x \; \frac{mcg/kg}{min} \; x \; \frac{pd}{1} = \frac{cc}{hr}$$

$$\frac{kg}{pd} \; x \; \frac{mg}{mcg} \; x \; \frac{min}{hr} \; x \; \frac{\boldsymbol{cc}}{\boldsymbol{mg}} \; x \; \frac{mcg/kg}{min} \; x \; \frac{pd}{1} = \frac{cc}{hr}$$

$$\frac{1\,kg}{2.2\,pd} \; x \; \frac{1\,mg}{1000\,mcg} \; x \; \frac{60\,min}{1\,hr} \; x \; \frac{250\,cc}{100\,mg} \; x \; \frac{5\,mcg/kg}{1\,min} \; x \; \frac{220\,pd}{1} =$$

$$\frac{1\,\cancel{kg}}{\underset{1}{\cancel{2.2\,pd}}} \; x \; \frac{1\,\cancel{mg}}{\underset{4}{\cancel{1000\,mcg}}} \; x \; \frac{60\,\cancel{min}}{1\,hr} \; x \; \frac{\overset{1}{\cancel{250}}\,cc}{100\,\cancel{mg}} \; x \; \frac{5\,\cancel{mcg/kg}}{1\,\cancel{min}} \; x \; \frac{\overset{100}{\cancel{220\,pd}}}{1} =$$

$$\frac{1}{\underset{1}{4}} \times \frac{\overset{15}{\cancel{60}}}{1\ hr} \times \frac{1\ cc}{\underset{1}{\cancel{100\ mg}}} \times \frac{\overset{1}{\cancel{100}}}{1} \times \frac{5}{1} = \frac{75\ cc}{hr}$$

Problem 16.

A new order arrives stating that O should be reduced to 3 mcg/kg/hour, if the patient's blood pressure drops below 140/80. It does.
 At how many cc/hr would the IV now run?

Notice that the above problem actually broke down to 15 x 5 cc/hr, the 5 coming from the mcg/kg/min segment of the problem. Substituting a 3 for the 5 makes the number portion 15 x 3, or 45cc/hr. The problem would look like this in the short form:

$$\frac{1\;\cancel{kg}}{\underset{1}{\cancel{2.2\;pd}}} \; x \; \frac{1\;\cancel{mg}}{\underset{4}{\cancel{1000\;mcg}}} \; x \; \frac{60\;\cancel{min}}{1\;hr} \; x \; \frac{\overset{1}{\cancel{250}}\;cc}{100\;\cancel{mg}} \; x \; \frac{3\;\cancel{mcg/kg}}{1\;\cancel{min}} \; x \; \frac{\overset{100}{\cancel{220\;pd}}}{1} =$$

$$\frac{1}{\underset{1}{4}} \; x \; \frac{\overset{15}{\cancel{60}}}{1\;hr} \; x \; \frac{1\;cc}{\underset{1}{\cancel{100\;mg}}} \; x \; \frac{\cancel{100}}{1} \; x \; \frac{3}{1} = \frac{45\;cc}{hr}$$

Or like this in the more complete form:

$$\frac{kg}{pd} \; x \; \frac{mg}{mcg} \; x \; \frac{min}{hr} \; x \; \frac{mg}{cc} \; x \; \frac{mcg/kg}{min} \; x \; \frac{pd}{1} = \frac{cc}{hr}$$

$$\frac{kg}{pd} \; x \; \frac{mg}{mcg} \; x \; \frac{min}{hr} \; x \; \frac{cc}{mg} \; x \; \frac{mcg/kg}{min} \; x \; \frac{pd}{1} = \frac{cc}{hr}$$

$$\frac{1\;kg}{2.2\;pd} \; x \; \frac{1\;mg}{1000\;mcg} \; x \; \frac{60\;min}{1\;hr} \; x \; \frac{250\;cc}{100\;mg} \; x \; \frac{3\;mcg/kg}{1\;min} \; x \; \frac{220\;pd}{1} =$$

$$\frac{1\ \cancel{kg}}{\underset{1}{2.2\ \cancel{pd}}} \times \frac{1\ \cancel{mg}}{\underset{4}{1000\ \cancel{mcg}}} \times \frac{60\ \cancel{min}}{1\ hr} \times \frac{\overset{1}{\cancel{250}}\ cc}{100\ \cancel{mg}} \times \frac{3\ \cancel{mcg/kg}}{1\ \cancel{min}} \times \frac{\overset{100}{\cancel{220}}\ \cancel{pd}}{1} =$$

$$\frac{1}{\underset{1}{4}} \times \frac{\overset{15}{\cancel{60}}}{1\ hr} \times \frac{1\ cc}{\underset{1}{\cancel{100}\ \cancel{mg}}} \times \frac{\overset{1}{\cancel{100}}}{1} \times \frac{3}{1} = \frac{45\ cc}{hr}$$

Problem 17.

A patient weighing 150 kg is to have an IV of 50 mg Q in 250 cc NS to run at 0.5 mcg/kg/min. At how many cc/hour would this IV run?

Conversion factor? = yes, 1 mg = 1000 mcg, 1 hr = 60 min
Available dose = 50 mg/250 cc
Needed dose = 0.5 mcg/kg/min for pt 150 kg
Unknown = cc/hr

$$\frac{mg}{mcg} \times \frac{hr}{min} \times \frac{mg}{cc} \times \frac{mcg/kg}{min} \times \frac{kg}{1} = \frac{cc}{hr}$$

$$\frac{mg}{mcg} \times \frac{\boldsymbol{min}}{\boldsymbol{hr}} \times \frac{\boldsymbol{cc}}{\boldsymbol{mg}} \times \frac{mcg/kg}{min} \times \frac{kg}{1} = \frac{cc}{hr}$$

$$\frac{1\ mg}{1000\ mcg} \times \frac{60\ min}{1\ hr} \times \frac{250\ cc}{50\ mg} \times \frac{0.5\ mcg/kg}{1\ min} \times \frac{150\ kg}{1} = \frac{cc}{hr}$$

$$\frac{1\ \cancel{mg}}{\underset{4}{\cancel{1000\ mcg}}} \times \frac{60\ \cancel{min}}{1\ hr} \times \frac{\overset{1}{\cancel{250}}\ cc}{\underset{1}{\cancel{50\ mg}}} \times \frac{0.5\ \cancel{mcg/kg}}{1\ \cancel{min}} \times \frac{\overset{3}{\cancel{150\ kg}}}{1} = \frac{cc}{hr}$$

$$\frac{1}{\underset{1}{4}} \times \frac{\overset{15}{\cancel{60}}}{1\ hr} \times \frac{1\ cc}{1} \times \frac{0.5}{1} \times \frac{3}{1} = \frac{22.5\ cc}{hr}$$

Problem 18.

A new order arrives stating that Q should run at 1.0 mcg/kg/min if the patient's intracranial pressure rises above a reading of 12. It rises.
 Now how many cc/hr would the IV run?

As with the problems 15 and 16, this is also easily solvable. The equation in Problem 17 breaks down to 15 x 3 x 0.5, with 0.5 being the mcg/kg/min needed. Problem 18 is therefore 15 x 3 x 1, with 1 being the mcg/kg/min needed in this problem. Except for the mcg value, nothing else in the problem has changed. Substituting 1 for 0.5 yields the new result. Notice that all we have done is doubled the IV rate. Every factor has remained the same with the exception of the micrograms desired per hour, which in this case is twice what it was for the previous problem.

To show that the result is indeed what is described here, the short form of the problem would look like this:

$$\frac{1 \, mg}{\underset{4}{1000 \, mcg}} \; x \; \frac{60 \, min}{1 \, hr} \; x \; \frac{\overset{1}{250 \, cc}}{\underset{1}{50 \, mg}} \; x \; \frac{1 \, mcg/kg}{1 \, min} \; x \; \frac{\overset{3}{150 \, kg}}{1} = \frac{cc}{hr}$$

$$\frac{1}{\underset{1}{4}} \; x \; \frac{\overset{15}{60}}{1 \, hr} \; x \; \frac{1 \, cc}{1} \; x \; \frac{1}{1} \; x \; \frac{3}{1} = \frac{45cc}{1 \, hr}$$

About the Author

The author is a former nurse with other 40 years of experience in health care, most of them practiced in surgical intensive care. Most of these were spent at the University of Minnesota Hospitals and Clinics, a transplant center; although part of them were spent at St Lukes-Roosevelt Hospital, a trauma center in New York City.

Because mathematics came with difficulty to Ms Wilson as a child and young adult, she has maintained a life-long interest in it and continues to work mathematics problems as a "hobby." As she reports, "If you look at math problems as just a kind of 'puzzle,' they seem less intimidating to you, and the more of them you work, the better you understand how the whole thing works."

www.ingramcontent.com/pod-product-compliance
Lightning Source LLC
Chambersburg PA
CBHW081141170526
45165CB00008B/2749